Larry Burkett's
ALL ABOUT TIME

Discovering How the Calendar Affects You

Written by
KEVIN MILLER

Illustrated by
GARY LOCKE

A Faith Building Guide can be found on page 32.

Faith Kidz is an imprint of Cook Communications Ministries,
Colorado Springs, Colorado 80918
Cook Communications, Paris, Ontario
Kingsway Communications, Eastbourne, England

ALL ABOUT TIME ©2003 by Burkett & Kids, LLC

All rights reserved. No part of this publication may be reproduced without written
permission, except for brief quotations in books and critical reviews. For information, write Cook
Communications Ministries, 4050 Lee Vance View, Colorado Springs, CO 80918.

First printing, 2003
Printed in Singapore.
1 2 3 4 5 6 7 8 9 10 Printing/Year 07 06 05 04 03

All Scripture quotations, unless otherwise indicated, are taken from the HOLY BIBLE NEW INTERNATIONAL
VERSION® (NIV®) Copyright ©1995, 1996, 1998 by International Bible Society. Used
by permission of Zondervan Publishing House. All rights reserved.

Library of Congress Cataloging-in-Publication Data

Burkett, Larry.
    All about time / Larry Burkett ; illustrated by Gary Locke.
       p. cm.
Summary: Discusses the concept of time from a Christian perspective,
examining how time is measured and how it should be spent.
   ISBN 0-7814-3788-1
    1. Calendar—Juvenile literature. 2. Time—Juvenile literature.
3. Time measurements—Juvenile literature. [1. Time—Religious
I. aspects—Christianity. 2. Time measurements. 3. Christian life.]
Locke, Gary, ill. II. Title.
   CE13 .B87 2002
   529.3—dc21
                       2001006709

Stewardship for the Family
Executive Producer: Allen Burkett

Lightwave Publishing
Concept Direction: Rick Osborne

Cook Communications Ministries
Senior Editor: Heather Gemmen
Designer: Keith Sherrer, iDesignEtc.
Design Manager: Jeffrey P. Barnes

CONTENTS

In the Beginning 4

What Is Time? 6

Years, Months, & Seasons 8

Weeks, Days, Hours, Minutes, & Seconds 10

Calendars 12

Clocks & Watches 14

Time Zones 16

Is Time Travel Possible? 18

Can Anyone Predict the Future? 20

What Are the "End Times"? 22

What Is Eternity? 24

Why Do We Grow Old? 26

Managing Your Seconds, Minutes, & Hours 28

Managing Your Weeks, Months, & Years 30

Faith Building Guide 32

IN THE BEGINNING

Passing time," "wasting time," "in the nick of time," "time out," "time flies," "time's up," "on time." We sure have a lot of sayings about time. But why? Perhaps it's because way deep down, we know time is all we have. Once our earthly time is up, it's lights out, game over, do not insert quarters for extra play.

Like everything else God has created, our lives each have a beginning, and one day they will all come to an end. It's not a nice thing to think about, but knowing our lives will end encourages us to make the most of the time we have. That means getting in as much roller-blading, swimming, video game playing, and TV watching as possible, right? Not exactly. Those things are fun, and God likes us to have fun. But he also wants us to use our time wisely, doing things—like helping others and sharing our faith—that will last for eternity. Don't know what eternity is? Don't worry, it's a difficult topic, and we'll get to it later. Just know that using your time to serve God and others is the best way to live. And planning how to use your time is the best way to make that happen.

HOW OLD IS GOD?

We said everything had a beginning, but what about God? Did he have a beginning? If so, isn't he getting a little old to be running the universe? No way! God has always existed, and he will always exist. "'I am the Alpha and the Omega,' says the Lord God, 'who is, and who was, and

ALL ABOUT TIME

who is to come, the Almighty'" (Revelation 1:8). We may not understand this, but we can trust God when he says this is true.

WHERE DID TIME COME FROM?

Time was created by God. The Bible tells us that to God, a thousand years seems like a day, and a day seems like a thousand years (2 Peter 3:8). God isn't affected by time like we are, because he created it, and he controls it. God lives outside of time. That means the past, present, and future are all the same to him. He can see them all at once.

Isn't he getting a little old to be running the universe?

WHAT IS TIME?

Can you see time, hold it, smell it, or taste it? Not really. Time isn't a physical object like a pickle or a platypus. It's more like love or magnetism. We can't really see any of these things, but we can see and measure their effects. Check this out for yourself. Grab a photo album and look at a picture of yourself as a baby. Now go look in the mirror. See a difference? That's what time does to you. The longer you're in it, the more you change.

But how does time work? Where does it come from, and where does it go? Does it ever stop, speed up, or slow down? (It seemed like it stopped once. Read Joshua 10:12–14 to find out more.) These questions are difficult to answer. Humankind has puzzled over them for thousands of years. During that time, people have come up with all sorts of ways to explain time.

Have you ever heard the saying "History never repeats itself"? Some people haven't. They think time goes around in a big circle, and everything eventually repeats itself, including our lives. But that simply isn't true. The Bible says we only live and die once (Hebrews 9:27). So your life is moving forward in a straight line even though it may not feel that way when your alarm clock goes off at 7:00 A.M.—again!

Today, most people believe time just keeps moving forward with one thing happening after the other. We can't move backward, and we can't jump

ahead of the present. (More about time travel later.) The time behind us is called "history," and the time ahead of us is called "future." The Bible definitely confirms this view. It says time has a beginning, when God created the heavens and the earth (Genesis 1:1), and a definite end, when Jesus returns. At that time the old heaven and earth will pass away, and a new heaven and earth will be created in their place (2 Peter 3:13; Revelation 21).

FACTOID

What If We Didn't Keep Track of Time?

If we threw out all of our watches, calendars, and clocks, would we be able to stop time and stay young forever? That's like asking if we stopped measuring the temperature, would it always stay the same? Just as thermometers don't control the temperature, clocks, watches, and calendars don't control time; they simply measure it. And time continues to pass whether we measure it or not. So hold on to your watch!

There's no stopping time!

YEARS, MONTHS, & SEASONS

Since there's no stopping time, we might as well look at some ways of measuring it. We'll start with big chunks of time: years, months, and seasons.

YEARS

Have you ever wondered why a year is 365 days long? That's because a year measures the time it takes the earth to revolve around the sun. It actually takes 365.2422 days, but we add an extra day to the calendar every four years (February 29) to make up for the difference. We call those years "leap years."

Do you know what ten years are called? A decade. **How about 100 years?** A century. **One thousand years?** A millennium.

MONTHS

Every year is made up of 12 months. **Can you guess what a month measures?** The time it takes the moon to revolve around the earth.

The moon takes just over 29 days to go around the earth. But the moon's cycle isn't completely regular, so our calendar is divided up into 12 unevenly numbered months.

SEASONS

A season is a period of time that is marked by a change in weather and temperature. In North America, we have four seasons: winter, spring, summer, and fall. But some places, especially near the equator, have only two real seasons: the wet season and the dry season.

ALL ABOUT TIME

The seasons change because as the earth orbits the sun, it is tilted to one side on its *axis* (an imaginary pole running through the center of the earth from north to south). At some times of the year, the Northern Hemisphere, where we live, is tilted toward the sun, giving us more sunlight and longer days.

This is when we get our summer and the people in the Southern Hemisphere get their winter. But as the earth continues around the sun, we begin to tilt away from it. Our weather becomes colder, turning to winter, and the days become shorter. Meanwhile, the people in the Southern Hemisphere are pulling out their shorts and sunscreen.

FACTOID

Do you know how many days are in each MONTH?

Here's a rhyme to help you remember:

Thirty days hath September,
April, June, and November;
All the rest have 31,
excepting February alone:
Which hath but 28, in fine,
till leap year gives it 29.

Thirty days hath September, April, June...

WEEKS, DAYS, HOURS, MINUTES, & SECONDS

WEEKS

A week is seven days long, but it hasn't always been this way. The ancient Greeks lived by a ten-day week, and the early Romans used a nine-day week. We got our week from the Jews. They used a seven-day week because God told them this was how long it took him to create the world—six days of work and one day of rest—and he wanted them to work on a similar schedule (Exodus 20:9–11).

DAYS

A day measures the time it takes the earth to rotate on its axis. Remember that invisible line running through the earth's center? Not only does the earth spin around the sun, it also spins completely around on its axis every 24 hours. It's a wonder we don't get dizzy! When our part of the earth is facing the sun, we get day. When it is turned away, we get night.

HOURS, MINUTES, & SECONDS

Why is a day divided into 24 hours? No real reason. We just started doing it this way, and the habit has stuck. Can you imagine how things would be different if we only had 12 hours in a day? How about if we had 48 hours in a day?

ALL ABOUT TIME

Minutes are 1/60 of an hour. And seconds? They're 1/60 of a minute. Seconds can be broken down into smaller segments as well, such as *nanoseconds*, which are one billionth of a second. But only people like scientists and Olympic bobsledders who need extremely accurate measurements use these smaller units.

It's a wonder we don't get dizzy!

FACTOID

If a day is 24 hours long, why do our clocks only go to 12?

That's because we simply repeat the same 12-hour cycle each day. The first cycle, which lasts from midnight to 11:59 in the morning, is called "A.M." This is short for *ante meridiem*, which means "before noon" in Latin. The second cycle, which lasts from noon to 11:59 at night, is called "P.M." This is short for *post meridiem* or "after noon."

Some people, such as doctors and soldiers, use a 24-hour clock. On this clock, midnight is called 0:00, and time advances from there until we reach 23:59, which is the same as 11:59 P.M. Therefore, 1:00 P.M. becomes 13:00 (13 hours after midnight) and 6:00 P.M. becomes 18:00 (18 hours after midnight). But you don't say "thirteen o'clock" or "eighteen o'clock." You say "thirteen hundred hours" or "eighteen hundred hours." That's a military tradition.

CALENDARS

HOW WE GOT OUR CALENDAR

Our calendar came from the Romans, and it took thousands of years to develop. Here are some key steps along the journey.

Like other ancient peoples', the first Roman calendar was based on the moon's cycles. Their calendar had 10 months that added up to a 304-day year. But with 61.25 days missing, the priests had to delay announcing the New Year by that many days each winter so the farmers could plant their crops on time.

In 712 B.C., Emperor Numa tried to put an end to this nonsense by adding two more months to the calendar. This was a move in the right direction, but the calendar was still short by just over 11 days.

Julius Caesar thought he would one-up Numa in 46 B.C. by getting the calendar back on track. He changed the number of days in each of the 12 months: 30, 31, or 29 days, with an extra day in February on every fourth year. This was called a *leap year*.

He also named the seventh month after himself (Julius = July).

Not to be outdone, Augustus Caesar dove into the calendar problem as well. He also stamped his name on it, replacing "Sextilis" with "Augustus" (August).

He even took a day from "Februarius," which had 29 days, and added it to Augustus so it would have as many days as Julius.

While the emperors were squabbling over who had the longest month, their calendar was still losing time—11 minutes and 14 seconds per year, to be precise. This wasn't a problem for about 1,500 years. By then, the calendar was about 10 days off.

ALL ABOUT TIME

To fix this problem, on October 4, 1582 Pope Gregory XIII simply cut out 10 days from the calendar that year. He also changed the leap year rule so the problem wouldn't happen again. This calendar, called the Gregorian calendar, is the same one we use today.

Believe it or not, it is still out by 26 seconds per year. That means by 4905, we'll have to cut out one extra day. But none of us will be around to worry about that.

Wow! Those guys were smart!

CLOCKS & WATCHES

EARLY CLOCKS

The first clocks were probably just sticks in the ground that used the stick's shadow to track the sun's movement across the sky. These were called "shadow clocks." One kind of shadow clock you may have heard of is a *sundial*. But shadow clocks had some serious problems: They didn't work when the sun wasn't shining, and they were only good for measuring large segments of time.

Water clocks helped solve some of these problems. For one thing, they allowed people to keep time at night. Water clocks kept time by dripping water out of a tiny hole in the bottom of a container at a controlled rate so that people knew exactly how much water the container lost in one hour. Lines on the container marked off each hour. Later, they became much more complicated. But, like shadow clocks, water clocks had their problems. The water often froze, evaporated, or got so dirty it didn't run through properly.

BIG IMPROVEMENTS

Mechanical clocks, which were invented in the 14th century, were a major leap forward. The first ones were like today's cuckoo clocks. They kept time by slowly dropping a weight, which moved gears inside the clock. But they were still only accurate to within two hours.

Pendulum clocks, such as grandfather clocks, helped solve the accuracy problem. They kept time by swinging a pendulum back and forth at a regular speed. The first ones were a little rough.

ALL ABOUT TIME

But by 1900, pendulum clocks were accurate to within 1/100 of a second.

By the 1920s, people realized they could keep even better time by using vibrating quartz crystals instead of pendulums. Now they could keep accurate time to within 2/1000 of a second. This record was improved even more in 1956 when the atomic clock was invented. This clock is so accurate it will only lose one second every 30,000 years!

What time is it?

FACTOID

Watches weren't always worn on the wrist.

The first watches hung from chains around people's necks or from their pockets. But in 1880, German Navy officers started wearing watches on their wrists because it was too dangerous for them to take their hands off the railing in rough seas and pull out their watches. This habit was so convenient, it soon caught on everywhere. By the early 20th century, everyone was doing it.

TIME ZONES

Have you ever wondered why the time on the other side of the country is different than the time where you live? Read on.

THE TIME PROBLEM

Up until about 150 years ago, most people set their watches by the sun. When the sun was directly overhead, they knew it was noon, and they set their watch or clock. But this wasn't the best way of doing things. For one thing, because of the earth's rotation, the time when the sun is directly overhead is different depending on where you are. For instance, the sun appears directly overhead two hours earlier in Iowa than in California. That means if someone from California was scheduled to arrive in Iowa at noon on a train, he would be two hours late because his watch would be set to California time. Confusing? It certainly was.

THE TIME ZONE SOLUTION

Those who were mainly affected by this time problem were the railroad companies, because people were constantly missing trains or trains showed up late due to time mix-ups. The railways tried all kinds of things to straighten out this

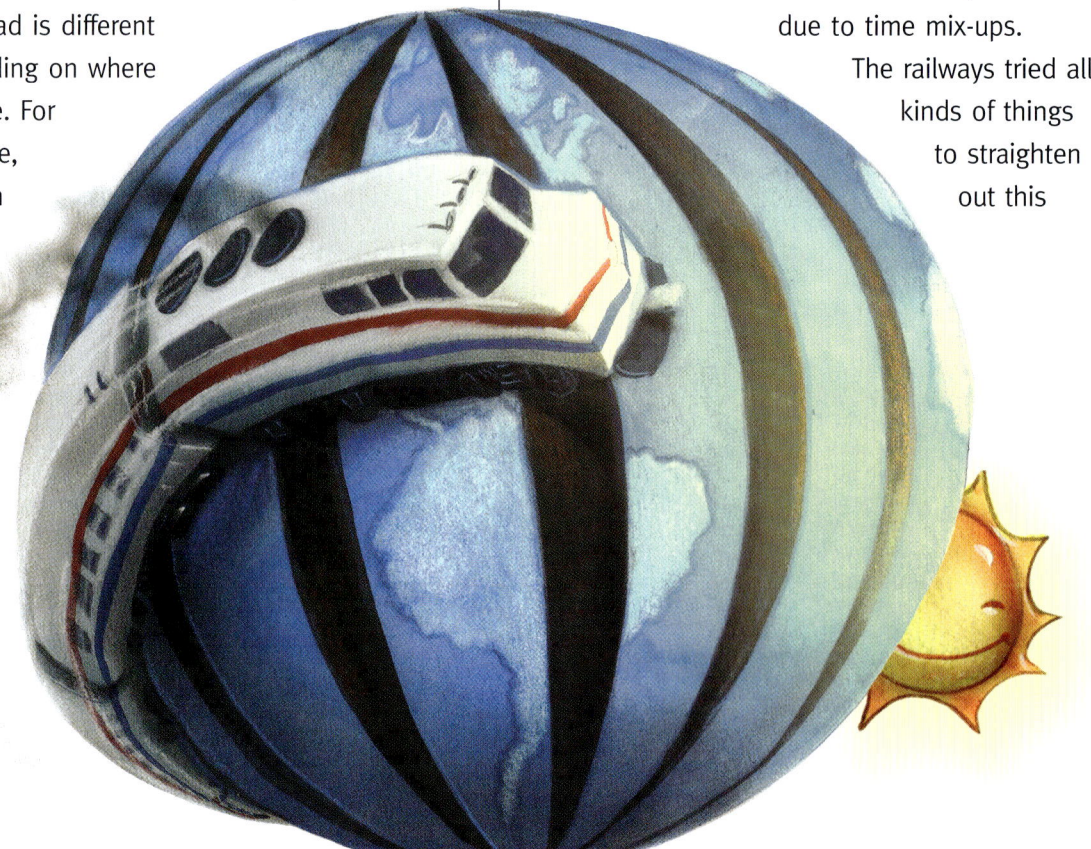

ALL ABOUT TIME

mess, but nothing worked as well as an idea suggested by Charles Dowd, a schoolteacher. He said the United States should be divided into four time zones, running from east to west. The difference between each time zone would be one hour. This way, everyone in the same time zone could set his or her watch to the same time. No more squinting at the sun, no more missed trains.

On November 18, 1883 the railroad companies began using this new system. It worked great—so great that the rest of the world adopted the system one year later. Today, the entire world is divided into 24 time zones, beginning in Greenwich, England. The continuous United States has four time zones: Eastern, Central, Mountain, and Pacific. What time zone do you live in?

What "zone" are you in?

FACTOID

Dividing the world into time zones worked well for setting train schedules. But it also created a new problem: now daylight came too early in the spring, and in the fall, it came too late. This problem was corrected with Daylight Savings Time (DST). This invention helps us make better use of daylight by setting our clocks one hour forward in the spring and one hour back in the fall. Remember: "Spring forward; fall back."

IS TIME TRAVEL POSSIBLE?

Have you ever wished you could travel back in time to meet someone famous? Could this really be possible? Some scientists think so. But most agree that if time travel is possible, we won't be doing it any time soon.

The main barrier to time travel is the enormous amount of energy required. To travel through time, we would have to move faster than light, which travels at 186,000 miles per second. But the closer we get to the speed of light, the heavier we get, so the more energy we need to move. But the more energy we add, the heavier we get. As you can see, it would just never work.

There may be shortcuts through space, such as *wormholes*, which would allow us to travel back in time without having to go faster than light. But no one really knows if wormholes exist—except in science fiction books and movies—and it would still take enormous amounts of power to open one up if they did.

ALL ABOUT TIME

And since no "time tourists" have come back from the future to tell us what it's like, it's unlikely that any of us will be touring through time either.

So you want to be a time tourist?

FACTOID

One way you can travel through time today is by crossing time zones. For example, if you're on a fast jet, it's possible to leave New York at 8:00 and get to Los Angeles by 7:00, one hour earlier. Or, if you fly east across the International Date Line, which runs from north to south just off the coast of Alaska, you can go from "tomorrow" to "today." Turn around, and you can go "back to the future."

But be careful! All that time travelling can catch up to you in the form of *jet lag*. We get jet lag when our bodies are running on one time, but the clocks are running on another. For example, if your plane leaves Los Angeles for London at 4:00 P.M. and gets you there eight hours later, your body is telling you it's midnight—time for bed. But in England, which is six time zones ahead, the clocks are telling you it's 8:00 A.M.—time to get up! It takes your body a long time to adjust to such drastic changes. The groggy feeling you get during this adjustment period is called jet lag.

CAN ANYONE PREDICT THE FUTURE?

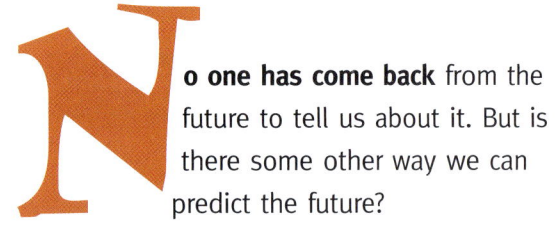

No one has come back from the future to tell us about it. But is there some other way we can predict the future?

FUTURISTS AND FAKERS

Many people make a living trying to predict the future of the economy or the stock market. They do this by looking at the way things are right now and how they were in the past and then trying to figure out if things will get better or worse down the road.

These people, called *futurists*, don't claim to be prophets, just good guessers.

Other people try to predict the future by reading people's palms, looking at horoscopes, reading *tarot* cards, or looking into crystal balls. But God forbids this kind of "fortune telling" (Deuteronomy 18:10), because he wants us to trust our future to him, not some prediction based on the creases in our palms. These methods of reading the future are usually inaccurate anyway. And if they do seem accurate, it is usually due to chance or wishful thinking. It's better if we just trust God to tell us as much about the future as we need to know.

ONLY GOD KNOWS

In truth, only God knows the future, because, as we learned in chapter one, he can see the present,

ALL ABOUT TIME

future, and past all at once. However, God has let some special people have a peek into the future. These people are called *prophets*. God allowed prophets to see into the future so they could warn others of what was about to happen. At other times, he showed them the future so they would encourage people to follow God and do what's right.

Some people think there are still prophets around today, but the Bible isn't exactly clear about this. Whatever the case, we do know that the future is in God's hands, and we can trust him to look out for us.

The future is in God's hands.

FACTOID

One way we know the Bible is true is because many of its prophecies have been fulfilled. A prophecy is something spoken or written down by a prophet that comes true in the future. Check out some of these fulfilled prophecies about Jesus.

More than 30 prophecies were made about Jesus in the Bible, and all of them have come true!

PROPHECY

Jesus will be born to a virgin
  R: Isaiah 7:14 **F:** Matthew 1:18–25; Luke 1:26–35

Jesus will be betrayed by a friend
  R: Psalm 41:9 **F:** John 13:18, 21, 26–27

Jesus will perform miracles
  R: Isaiah 35:5–6 **F:** Matthew 11:3–6; John 11:47

Jesus' bones won't be broken
  R: Psalm 34:20 **F:** John 19:31–37

Jesus will rise from the dead
  R: Psalm 16:10 **F:** Acts 13:35–37

R = *Reference*   **F** = *Fulfillment*

WHAT ARE THE "END TIMES"?

Something else that is prophesied or predicted in the Bible is a period of time called the "end times" or the "last days" when Jesus will return to earth. There are often a lot of misunderstandings about this event, even among adults. So if you're confused about it, you're in good company.

Part of this confusion stems from the fact that most prophecies about the end times are full of wild visions with dragons, dark riders, and all sorts of strange creatures and events. Sometimes the meaning of the visions is given; sometimes it is not. There are enough missing pieces to keep us guessing until Jesus returns. What we do know, however, is that Jesus will return one day, and he will come when we least expect it (Matthew 24:39–42; 1 Thessalonians 5:2).

He said no one knows the day or hour except God (Mark 13:32). But he also said he wouldn't return until the gospel had been preached to every nation (Revelation 14:6). The first time Jesus came, it was to give us salvation—to take the punishment for our sins so we can have a relationship with God. But when he returns, he is coming to judge

ALL ABOUT TIME

all people on earth according to how they responded to him. If they accepted his gift of salvation, they will be saved. Jesus will take them away to live with him forever in a new heaven and a new earth where there will be no more tears, crying, death, or pain (Revelation 21:4). But whoever rejects Jesus will have to be separated from him forever.

Jesus will take them away to live with him forever!

FACTOID

If you're already a Christian, that's great! But if you're not a Christian, you can become one right now. Get a Bible and read John 3:16 and Romans 10:9–10. Then pray the following prayer: "Dear God, thank you for forgiving me for the things I've done wrong. I know Jesus died for my sins, and I believe you raised him from the dead. I want Jesus as my Lord and Master. Please take control of my life and help me become the good person you made me to be. Amen."

Once you've prayed this prayer, make sure you tell your parents or a pastor about it so they can help you begin your new life as a Christian.

WHAT IS ETERNITY?

From here to eternity." "To infinity, and beyond!" These sorts of phrases often find their way into popular entertainment and everyday conversations. But what exactly do they mean? And how is infinity different from eternity?

ETERNITY VS. INFINITY

Both *infinity* and *eternity* were words originally coined to describe God. However, now we also use the term *infinity* in mathematics to describe a number that is so large it is uncountable.

While infinity and eternity are related, they don't mean quite the same thing. Despite what mathematicians might say, infinity is a quality that only God has. To say God is infinite is to say he is not bound by time or space. Like we said in chapter one, God created time, and he lives outside of it. He also created space. And while he is here, there, and everywhere at once, he is also infinitely larger than our world or the universe in which we live.

Eternity, on the other hand, is a period of time without beginning or end. We can also say that God is eternal, because he doesn't have a beginning or end. He just is. That's it.

YOU CAN LIVE FOREVER TOO!

The good news is that God has offered to give us eternal life through Jesus. That means if you are a Christian, you will go to live with God forever after you die.

But don't be confused: eternal life is about more than just endless existence. Everyone is going to live forever, whether they spend eternity with God or not. The difference is, eternal life with God will be just as fun and exciting as it is long. People who aren't with God will spend the whole time wishing it would end. But it never will.

ALL ABOUT TIME

FACTOID

Here's a brain tickler for you:

Imagine a library with an infinite number of red books and an infinite number of green books. If you took away all of the green books, how many books would you have left?

Answer: an infinite amount. How can that be? Because you had an infinite number of both types of books. Infinity minus infinity is still infinity! Head hurting yet? That's just the sort of thing that happens when you start dealing with infinite numbers.

WHY DO WE GROW OLD?

Time is just like a hot tub: when you first get in you look fine, but after a while, your skin gets all swollen and wrinkled. Why is this?

SCIENTIFICALLY SPEAKING

Believe it or not, scientists don't actually know why we grow old. They have some ideas, and they've even managed to double the life of fruit flies in some experiments. (Good news if you're a fruit fly!) But they still don't know enough about the aging process to do much more than measure how quickly it happens.

One theory of aging is that our cells are programmed to reproduce themselves only a certain number of times before they die. When enough cells stop reproducing or dividing, our bodies break down and eventually stop working. This is called death. If we could reprogram our cells to reproduce more times, perhaps we could live longer. But scientists are still trying to figure out if and how they could do this in humans.

A BIBLICAL EXPLANATION

The Bible tells us that we are all *mortal*; we will all die one day (Genesis 6:3; Romans 8:11). Things weren't this way in the beginning. God originally created us to live forever. But when Adam and Eve sinned, death came into the world. From that time on, humans have died. It is only after death that we will become *immortal* and live forever.

Way back in biblical history, humans lived much longer than they do now. Methuselah, the oldest

ALL ABOUT TIME

man in the Bible, lived 969 years! The aging process must have worked much slower back then. But as time wore on, people lived shorter and shorter lives. Today, the oldest person on record is 122 years. Most people in North America tend to live no longer than 79 years.

You're nothing but a spring chicken.

FACTOID

You may be an old timer to a fruit fly, which has a lifespan of only a few days. But to a Bristlecone pine, which can live up to 4,900 years, you're nothing but a spring chicken!

Check out the ages of these other plants and animals.

Animals	Maximum Ages
Galapagos tortoise	152
Indian elephant	78
American lobster	50
Bald eagle	44
Dog	29
Bullfrog	16
Earthworm	10

Plants	Maximum Ages
European larch	2,300
English oak	1,500
Sierra redwood	750
Common juniper	417

MANAGING YOUR SECONDS, MINUTES, & HOURS

Time is our greatest treasure. Once time passes, it's gone for good. Therefore, it's important to make the most of our time while we still have it by managing it wisely, starting with the little bits of time: hours, minutes, and seconds. Maybe the best way to explain how to do this is through the following story.

GET THE BIG ROCKS IN FIRST

A professor put a large glass container on the table in front of his class. He pulled out three large rocks and put them in the jar.

"Is this jar full?" he asked the class.

"Yes," someone replied.

The professor bent down and picked up some smaller rocks. He dumped them into the jar, and they filled the spaces around the big rocks.

"And now?"

"Yes?" someone said again, but this time he wasn't so sure.

Then the professor pulled out some even smaller rocks and sprinkled them inside the jar. They filled up the spaces between the large and medium-sized rocks.

"How about now?" he asked the class. But they were catching on.

"The spaces where the little rocks can fit are full," someone said.

The professor smiled, pulled out a bag of sand, and poured it into the jar until it was full.

"Now," he said. "Can anyone guess what this object lesson was supposed to teach?"

Can you guess?

What the professor told his class was that the big rocks stood for the most important things in our day, the medium rocks for things that are a bit less important, the small rocks for things less important still, and the sand for things that are least important of all. The lesson is that if we don't get the big rocks in first, the jar will get so full of smaller rocks

ALL ABOUT TIME

and sand that we won't be able to get them in later on.

Remember this story when you're planning your day and your life. Get the big rocks in first. These include things like working, studying, and spending time with God, friends, and family. But life isn't all about work and duty. Once those are done, there's nothing wrong with relaxing and having fun. The key is to manage your time so that you always have enough for both work and play.

FACTOID

Other Time Management Tips

Live one day at a time (Matthew 6:34): "Therefore do not worry about tomorrow, for tomorrow will worry about itself. Each day has enough trouble of its own."

- Make a list of the things you need to do at the beginning of each day.
- Do the most important things first.
- Never put *things* before *people*.
- Always put time with God at the top of your list!

Live one day at a time!

MANAGING YOUR WEEKS, MONTHS, & YEARS

TIME FOR EVERYTHING

One of the keys to managing your life and serving God most effectively is learning to find a balance between work and play, activity and rest, social time and "alone" time, family and friends, and so on. As the writer of Ecclesiastes says, "There is a time for everything, and a season for every activity under heaven" (Ecclesiastes 3:1).

A common mistake people make is to live their lives out of balance. They choose one thing and do it so much it starts to hurt everything else in their life. The thing they choose to do might be a good thing, but if they do it too much, it becomes a bad thing. For example, it's good for you to do your schoolwork. But if schoolwork is all you do and you never get any exercise or visit with your friends and family, you will be lonely, unhealthy, and unhappy. So you can see it's wise to work toward finding balance in your entire life. This includes a good mix of physical, mental, personal, and social activities.

REMEMBER ETERNITY

At the beginning of this book, we said the best way to spend your time is to use it to love and serve God and others. No matter how much money you make, how many people you meet, or how many things you buy, you can't take any of it with you when you die. Only what you do for God and others will last into eternity. In other words, what you *give* is most important, not what you *get* or what you own. And, best of all, God has promised to reward you if you spend your time obeying him.

So whatever you do, never forget that your actions will affect your life far beyond this world. If you're ever in doubt about whether

ALL ABOUT TIME

or not you should do something, ask yourself: How will this affect my life for eternity? Will it please God or not? Will it benefit myself and others or not? The choice should be clear after that.

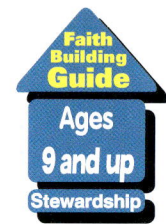

ALL ABOUT TIME

Spiritual Building Block: Stewardship

You can become better managers of time in the following ways:

Think About It: Next time you hear someone say, "What a waste of time," think about whether or not they are right. (Keep in mind that *waste* means: to use for no profit.) What kinds of activities waste time and what kinds of activities make the best use of time? Read Philippians 4:8–9 to find out how God wants you to spend your time.

Talk About It: Sit down with your parents or your siblings or a friend to evaluate how you use your time. How much time is used for work, for exercise, for prayer, for television, for reading, for complaining, for helping others, for sleeping? Talk about which activities you would like to make priorities and which you would like to cut down on. Remember, "even a child is known by his actions, by whether his conduct is pure and right" (Prov. 20:11).

Try It: Draw up a schedule (similar to the one below) that breaks down seven days into half-hour sections. For one week, keep careful track of everything you do, and then evaluate how you spent your week. Fill in the blanks of a new schedule for the following week with the goal of being wise in the use of your time.

	Sun	Mon	Tues	Wed	Thur	Fri	Sat
6:30-7:00							
7:00-7:30							
7:30-8:00							
8:00-8:30